Editor
Gisela Lee

Managing Editor
Karen Goldfluss, M.S. Ed.

Editor-in-Chief
Sharon Coan, M.S. Ed.

Cover Artist
Barb Lorseyedi

Art Coordinator
Kevin Barnes

Art Director
CJae Froshay

Imaging
Alfred Lau
James Edward Grace

Product Manager
Phil Garcia

Publisher
Mary D. Smith, M.S. Ed.

Practice Makes Perfect

Division

GRADE 4

Author

Robert W. Smith

Teacher Created Resources

Teacher Created Resources, Inc.
6421 Industry Way
Westminster, CA 92683
www.teachercreated.com

ISBN-0-7439-3324-9

©2002 Teacher Created Resources, Inc.
Reprinted, 2006
Made in U.S.A.

Table of Contents

Introduction

The old adage "practice makes perfect" can really hold true for your child and his or her education. The more practice and exposure your child has with concepts being taught in school, the more success he or she is likely to find. For many parents, knowing how to help their children can be frustrating because the resources may not be readily available.

As a parent it is also difficult to know where to focus your efforts so that the extra practice your child receives at home supports what he or she is learning in school.

This book has been designed to help parents and teachers reinforce basic skills with their children. *Practice Makes Perfect* reviews basic math skills for children in the fourth grade. The math focus is on division. While it would be impossible to include all concepts taught in the fourth grade in this book, the following basic objectives are reinforced through practice exercises. These objectives support math standards established on a district, state, or national level. (Refer to the Table of Contents for the specific objectives of each practice page.)

- division facts to 20
- division tables
- word problems
- rules of divisibility

- dividing with money
- one-digit division with and without remainders
- two-digit division with and without remainders
- three-digit division

There are 36 practice pages organized sequentially, so children can build their knowledge from more basic skills to higher-level math skills. To correct the practice pages in this book, use the answer key provided on pages 47 and 48. Six practice tests follow the practice pages. These provide children with multiple-choice test items to help prepare them for standardized tests administered in schools. As children complete a problem, they fill in the correct letter among the answer choices. An optional "bubble-in" answer sheet has also been provided on page 46. This answer sheet is similar to those found on standardized tests. As your child completes each test, he or she can fill in the correct bubbles on the answer sheet.

How to Make the Most of This Book

Here are some useful ideas for optimizing the practice pages in this book:

- Set aside a specific place in your home to work on the practice pages. Keep it neat and tidy with materials on hand.

- Set up a certain time of day to work on the practice pages. This will establish consistency. An alternative is to look for times in your day or week that are less hectic and are more conducive to practicing skills.

- Keep all practice sessions with your child positive and constructive. If the mood becomes tense or you and your child are frustrated, set the book aside and look for another time to practice with your child.

- Help with instructions if necessary. If your child is having difficulty understanding what to do or how to get started, work through the first problem with him or her.

- Review the work your child has done. This serves as reinforcement and provides further practice.

- Allow your child to use whatever writing instruments he or she prefers. For example, colored pencils can add variety and pleasure to drill work.

- Pay attention to the areas in which your child has the most difficulty. Provide extra guidance and exercises in those areas. Allowing children to use drawings and manipulatives, such as coins, tiles, game markers, or flash cards can help them grasp difficult concepts more easily.

- Look for ways to make real-life applications to the skills being reinforced.

Practice 1

The multiplication chart shown here can be used to find any basic multiplication or division fact until you have learned them all.

One of the best ways to learn the facts is to practice using the chart.

Columns

Rows	1	2	3	4	5	6	7	8	9	10	11	12
1	1	2	3	4	5	6	7	8	9	10	11	12
2	2	4	6	8	10	12	14	16	18	20	22	24
3	3	6	9	12	15	18	21	24	27	30	33	36
4	4	8	12	16	20	24	28	32	36	40	44	48
5	5	10	15	20	25	30	35	40	45	50	55	60
6	6	12	18	24	30	36	42	48	54	60	66	72
7	7	14	21	28	35	42	49	56	63	70	77	84
8	8	16	24	32	40	48	56	64	72	80	88	96
9	9	18	27	36	45	54	63	72	81	90	99	108
10	10	20	30	40	50	60	70	80	90	100	110	120
11	11	22	33	44	55	66	77	88	99	110	121	132
12	12	24	36	48	60	72	84	96	108	120	132	144

Read across for the Rows.

Read up or down for the Columns.

Note: To find how many times 8 divides into 56, run one finger across the 8 row until you come to 56 and run a finger up the column with 56 until you come to the top number which is 7. The answer is that 8 divides into 56 exactly 7 times.

Directions: Use the rows on the multiplication chart to help you find the missing numbers. (Go backwards.)

1. (24, 22, 20, 18, 16, 14, _____ , _____ , _____ , _____ , _____ , _____)
2. (36, 33, 30, 27, 24, 21, _____ , _____ , _____ , _____ , _____ , _____)
3. (48, 44, 40, 36, _____ , _____ , _____ , _____ , _____ , _____ , _____ , _____)
4. (60, 55, 50, 45, 40, _____ , _____ , _____ , _____ , _____ , _____ , _____)
5. (72, 66, 60, 54, 48, _____ , _____ , _____ , _____ , _____ , _____ , _____)
6. (84, 77, 70, 63, 56, _____ , _____ , _____ , _____ , _____ , _____ , _____)
7. (96, 88, 80, 72, 64, _____ , _____ , _____ , _____ , _____ , _____ , _____)
8. (144, 132, 120, 108, 96, 84, _____ , _____ , _____ , _____ , _____ , _____)
9. Which row has a zero in every number? _____

Practice 2

Directions: Use the columns on the multiplication/division chart to help you find the missing numbers. (Go backwards.)

1.	2.	3.	4.	5.
96	120	60	132	144
88	110	55	121	132
80	100	50	110	120
72	90	45	99	108
64	80	40	88	96
____	____	____	____	____
____	____	____	____	____
____	____	____	____	____
____	____	____	____	____
____	____	____	____	____
____	____	____	____	____
____	____	____	____	____

6. Which column has a zero in every number? _____

Directions: Use the multiplication/division chart to help you find the answers to these problems. The first two are done for you.

7. How many 8's can you subtract from 48?

   48 − 8 − 8 − 8 − 8 − 8 − 8 = 0

   You can subtract six 8's from 48.

8. How many 5's can you subtract from 30?

   30 − 5 − 5 − 5 − 5 − 5 − 5 = 0

   You can subtract six 5's from 30.

9. How many 6's can you subtract from 48?

   48 − 6 − 6 − 6 − 6 − 6 − 6 − 6 − 6 = 0

10. How many 7's can you subtract from 42?

    42 −

11. How many 3's can you subtract from 36?

    36 −

12. How many 10's can you subtract from 100?

    100 −

Practice 3

Directions: Do these problems. Use your multiplication/division chart if you are unsure of your division facts.

1. 5)25

2. 2)10

3. 4)12

4. 3)3

5. 1)9

6. 4)20

7. 6)54

8. 4)28

9. 3)15

10. 4)32

11. 5)5

12. 4)40

13. 5)55

14. 2)18

15. 1)6

16. 2)16

17. 2)10

18. 4)48

19. 3)9

20. 5)20

21. 4)24

22. 2)12

23. 4)16

24. 4)24

25. 3)21

26. 4)12

27. 4)28

28. 3)30

29. 2)16

30. 5)45

31. 4)44

32. 2)24

33. 3)27

34. 1)12

35. 4)48

36. 3)33

37. 2)20

38. 3)18

39. 5)50

40. 4)36

41. 3)36

42. 5)35

43. 2)22

44. 4)40

45. 3)24

46. 1)10

47. 4)32

48. 5)55

Practice 4

Directions: Do these problems. Use your multiplication/division chart if you are unsure of your division facts.

1. 6)24   2. 4)28   3. 9)27   4. 8)48

5. 6)48   6. 6)54   7. 9)54   8. 7)56

9. 8)64   10. 6)30   11. 7)21   12. 7)35

13. 7)14   14. 9)72   15. 6)18   16. 8)24

17. 9)99   18. 8)88   19. 6)36   20. 7)7

21. 7)49   22. 7)21   23. 6)60   24. 8)96

25. 7)14   26. 8)24   27. 8)72   28. 9)90

29. 6)18   30. 8)88   31. 7)28   32. 8)32

33. 6)24   34. 7)7   35. 9)81   36. 7)63

37. 8)40   38. 6)54   39. 7)70   40. 8)16

41. 7)77   42. 7)84   43. 6)42   44. 8)56

45. 6)12   46. 7)63   47. 6)72   48. 9)108

Practice 5

Directions: Do these problems. Use your multiplication/division chart if you are unsure of your division facts.

1. $10\overline{)10}$

2. $12\overline{)60}$

3. $11\overline{)55}$

4. $12\overline{)48}$

5. $12\overline{)24}$

6. $12\overline{)132}$

7. $11\overline{)88}$

8. $12\overline{)96}$

9. $12\overline{)72}$

10. $10\overline{)90}$

11. $10\overline{)80}$

12. $11\overline{)110}$

13. $11\overline{)66}$

14. $10\overline{)40}$

15. $10\overline{)20}$

16. $12\overline{)144}$

17. $11\overline{)55}$

18. $11\overline{)33}$

19. $12\overline{)84}$

20. $10\overline{)120}$

21. $12\overline{)36}$

22. $11\overline{)44}$

23. $11\overline{)99}$

24. $10\overline{)100}$

25. $12\overline{)60}$

26. $11\overline{)33}$

27. $11\overline{)22}$

28. $12\overline{)36}$

29. $10\overline{)80}$

30. $12\overline{)132}$

31. $11\overline{)66}$

32. $10\overline{)30}$

33. $11\overline{)11}$

34. $12\overline{)48}$

35. $10\overline{)60}$

36. $11\overline{)121}$

37. $11\overline{)55}$

38. $12\overline{)72}$

39. $12\overline{)12}$

40. $11\overline{)132}$

41. $10\overline{)100}$

42. $11\overline{)99}$

43. $10\overline{)110}$

44. $12\overline{)84}$

45. $10\overline{)40}$

46. $12\overline{)60}$

47. $10\overline{)50}$

48. $10\overline{)90}$

#3324 Practice Makes Perfect: Division © *Teacher Created Resources, Inc.*

Practice 6

Directions: Do these problems. Use your multiplication/division chart if you are unsure of your division facts.

1. $4\overline{)12}$
2. $3\overline{)18}$
3. $11\overline{)88}$
4. $9\overline{)36}$

5. $10\overline{)80}$
6. $9\overline{)27}$
7. $4\overline{)48}$
8. $7\overline{)35}$

9. $5\overline{)40}$
10. $12\overline{)60}$
11. $11\overline{)77}$
12. $12\overline{)108}$

13. $6\overline{)24}$
14. $2\overline{)18}$
15. $4\overline{)16}$
16. $7\overline{)21}$

17. $7\overline{)77}$
18. $3\overline{)24}$
19. $8\overline{)88}$
20. $11\overline{)121}$

21. $6\overline{)36}$
22. $9\overline{)18}$
23. $4\overline{)16}$
24. $9\overline{)45}$

25. $3\overline{)30}$
26. $9\overline{)36}$
27. $4\overline{)32}$
28. $10\overline{)30}$

29. $8\overline{)40}$
30. $10\overline{)50}$
31. $4\overline{)12}$
32. $7\overline{)84}$

33. $5\overline{)25}$
34. $3\overline{)36}$
35. $7\overline{)63}$
36. $4\overline{)28}$

37. $5\overline{)35}$
38. $4\overline{)20}$
39. $12\overline{)60}$
40. $5\overline{)50}$

41. $8\overline{)16}$
42. $9\overline{)90}$
43. $8\overline{)64}$
44. $10\overline{)120}$

45. $3\overline{)15}$
46. $5\overline{)15}$
47. $9\overline{)81}$
48. $7\overline{)49}$

Practice 7

Directions: Do these problems. Use your multiplication/division chart if you are unsure of your division facts.

1. $2\overline{)14}$ 2. $5\overline{)25}$ 3. $12\overline{)36}$ 4. $5\overline{)40}$

5. $11\overline{)77}$ 6. $6\overline{)18}$ 7. $9\overline{)81}$ 8. $4\overline{)20}$

9. $3\overline{)24}$ 10. $9\overline{)54}$ 11. $7\overline{)35}$ 12. $7\overline{)14}$

13. $2\overline{)24}$ 14. $4\overline{)12}$ 15. $6\overline{)24}$ 16. $12\overline{)48}$

17. $3\overline{)21}$ 18. $8\overline{)64}$ 19. $12\overline{)108}$ 20. $5\overline{)50}$

21. $4\overline{)24}$ 22. $8\overline{)32}$ 23. $5\overline{)45}$ 24. $9\overline{)99}$

25. $9\overline{)27}$ 26. $8\overline{)24}$ 27. $4\overline{)44}$ 28. $12\overline{)84}$

29. $8\overline{)56}$ 30. $9\overline{)90}$ 31. $5\overline{)55}$ 32. $12\overline{)36}$

33. $10\overline{)90}$ 34. $7\overline{)56}$ 35. $7\overline{)63}$ 36. $9\overline{)63}$

37. $9\overline{)45}$ 38. $8\overline{)80}$ 39. $4\overline{)16}$ 40. $6\overline{)30}$

41. $3\overline{)18}$ 42. $6\overline{)54}$ 43. $3\overline{)36}$ 44. $6\overline{)72}$

45. $3\overline{)15}$ 46. $4\overline{)36}$ 47. $9\overline{)18}$ 48. $10\overline{)70}$

Practice 8

Directions: Fill in the missing factors. Use your multiplication/division chart, if you are unsure of your facts.

1. $9 \times \underline{\quad} = 54$

2. $3 \times \underline{\quad} = 36$

3. $10 \times \underline{\quad} = 60$

4. $8 \times \underline{\quad} = 64$

5. $\underline{\quad} \times 9 = 81$

6. $6 \times \underline{\quad} = 72$

7. $9 \times \underline{\quad} = 45$

8. $4 \times \underline{\quad} = 48$

9. $\underline{\quad} \times 7 = 42$

10. $\underline{\quad} \times 7 = 63$

11. $\underline{\quad} \times 9 = 63$

12. $5 \times \underline{\quad} = 55$

13. $6 \times \underline{\quad} = 36$

14. $\underline{\quad} \times 4 = 36$

15. $\underline{\quad} \times 6 = 54$

16. $\underline{\quad} \times 9 = 99$

17. $7 \times \underline{\quad} = 49$

18. $9 \times \underline{\quad} = 63$

19. $42 \div 7 = \underline{\quad}$

20. $54 \div 9 = \underline{\quad}$

21. $54 \div 6 = \underline{\quad}$

22. $12 \div 3 = \underline{\quad}$

23. $88 \div 8 = \underline{\quad}$

24. $48 \div 12 = \underline{\quad}$

25. $56 \div 7 = \underline{\quad}$

26. $56 \div 8 = \underline{\quad}$

27. $96 \div 8 = \underline{\quad}$

28. $18 \div \underline{\quad} = 6$

29. $44 \div \underline{\quad} = 4$

30. $72 \div \underline{\quad} = 8$

31. $33 \div \underline{\quad} = 3$

32. $120 \div \underline{\quad} = 10$

33. $77 \div \underline{\quad} = 11$

34. $24 \div \underline{\quad} = 4$

35. $49 \div \underline{\quad} = 7$

36. $16 \div \underline{\quad} = 2$

Practice 9

Directions: Do these problems. Use your multiplication/division chart if you are unsure of your division facts. The first one is done for you.

1.
$$\begin{array}{r} 9\ \text{R3} \\ 4\overline{)39} \\ -36 \\ \hline 3 \end{array}$$

2. $6\overline{)57}$

3. $4\overline{)49}$

4. $5\overline{)56}$

5. $5\overline{)27}$

6. $6\overline{)67}$

7. $9\overline{)58}$

8. $8\overline{)95}$

9. $6\overline{)47}$

10. $8\overline{)69}$

11. $7\overline{)87}$

12. $9\overline{)51}$

13. $7\overline{)47}$

14. $4\overline{)85}$

15. $3\overline{)53}$

16. $4\overline{)49}$

17. $7\overline{)85}$

18. $4\overline{)59}$

19. $5\overline{)99}$

20. $4\overline{)61}$

21. $6\overline{)79}$

22. $5\overline{)76}$

23. $4\overline{)91}$

24. $3\overline{)73}$

Practice 10

Directions: Do these problems. Use your multiplication/division chart if you are unsure of your division facts. The first one is done for you.

1.
```
     11 R4
  7) 81
   - 7
     11
    - 7
      4
```

2. 9) 76

3. 4) 39

4. 7) 92

5. 8) 73

6. 9) 67

7. 5) 69

8. 8) 39

9. 4) 27

10. 9) 97

11. 4) 62

12. 12) 69

13. 10) 78

14. 3) 29

15. 4) 51

16. 6) 92

17. 6) 71

18. 4) 89

19. 11) 91

20. 12) 79

21. 8) 93

22. 7) 93

23. 10) 86

24. 4) 63

Practice 11

Directions: Do these problems. Use your multiplication/division chart if you are unsure of your division facts. The first two are done for you.

1.
```
      51
  3) 153
   - 15
      3
    - 3
      0
```

2.
```
      31
  4) 124
   - 12
      4
    - 4
      0
```

3. 5) 205

4. 4) 324

5. 3) 633

6. 5) 355

7. 6) 426

8. 4) 164

9. 9) 819

10. 3) 186

11. 3) 666

12. 7) 427

13. 6) 372

14. 8) 416

15. 3) 225

16. 3) 258

17. 4) 136

18. 5) 265

19. 5) 375

20. 4) 396

Practice 12

Directions: Do these problems. Use your multiplication/division chart if you are unsure of your division facts. The first two are done for you.

1.
```
      92
  6) 552
   - 54
     12
    -12
      0
```

2.
```
     123
  5) 615
   - 5
     11
    -10
     15
    -15
      0
```

3. 3) 648

4. 4) 512

5. 6) 822

6. 6) 444

7. 4) 568

8. 2) 492

9. 3) 756

10. 4) 936

11. 6) 978

12. 9) 909

13. 5) 600

14. 4) 824

15. 8) 848

Practice 13

Directions: Do these problems. Use your multiplication/division chart if you are unsure of your division facts. The first two are done for you.

1.
```
      103  R3
   4)415
    - 4
      15
    - 12
       3
```

2.
```
      77  R1
   3)232
    - 21
      22
    - 21
       1
```

3. 5)269

4. 3)319

5. 6)917

6. 9)263

7. 8)109

8. 2)327

9. 7)709

10. 4)419

11. 6)806

12. 4)818

13. 4)915

14. 4)729

15. 3)929

16. 5)951

Practice 14

Directions: Do these problems. Use your multiplication/division chart if you are unsure of your division facts. The first two are done for you.

1.
```
      122  R5
   6)737
    - 6
      13
    - 12
      17
    - 12
       5
```

2.
```
      175  R2
   3)527
    - 3
      22
    - 21
      17
    - 15
       2
```

3. 4)293

4. 4)531

5. 5)617

6. 2)193

7. 6)437

8. 8)239

9. 4)857

10. 6)273

11. 3)313

12. 7)629

13. 4)909

14. 3)733

15. 5)991

16. 6)185

Practice 15

Directions: Do these problems. Use your multiplication/division chart if you are unsure of your division facts. The first two are done for you.

1.
$$\begin{array}{r} 716 \\ 4\overline{)2864} \\ -28 \\ \hline 6 \\ -4 \\ \hline 24 \\ -24 \\ \hline 0 \end{array}$$

2.
$$\begin{array}{r} 515 \\ 5\overline{)2575} \\ -25 \\ \hline 7 \\ -5 \\ \hline 25 \\ -25 \\ \hline 0 \end{array}$$

3. $3\overline{)2409}$

4. $4\overline{)3216}$

5. $3\overline{)1809}$

6. $6\overline{)3642}$

7. $7\overline{)3570}$

8. $9\overline{)9909}$

9. $7\overline{)5607}$

10. $6\overline{)5046}$

11. $8\overline{)6512}$

12. $4\overline{)6588}$

13. $3\overline{)4146}$

14. $9\overline{)3222}$

15. $8\overline{)8808}$

Practice 16

Directions: Do these problems. Use your multiplication/division chart if you are unsure of your division facts. The first two are done for you.

1.
```
      590 R3
  4 )2363
   - 20
      36
    - 36
       3
```

2.
```
      717 R1
  2 )1435
   - 14
       3
     - 2
      15
    - 14
       1
```

3. 5)2761

4. 3)1927

5. 4)3317

6. 7)3828

7. 8)6019

8. 6)7133

9. 5)4218

10. 3)2831

11. 9)8819

12. 4)2917

13. 9)1919

14. 7)2333

15. 6)6211

Practice 17

Directions: Do these problems. Use your multiplication/division chart if you are unsure of your division facts. The first two are done for you.

1.
```
      645 R2
  3)1937
  - 18
    13
  - 12
    17
  - 15
     2
```

2.
```
     1300 R3
  4)5203
  - 4
    12
  - 12
     3
```

3. $6\overline{)1919}$

4. $5\overline{)3854}$

5. $9\overline{)7787}$

6. $4\overline{)5995}$

7. $6\overline{)1966}$

8. $8\overline{)8778}$

9. $5\overline{)2002}$

10. $8\overline{)2407}$

11. $6\overline{)1339}$

12. $3\overline{)2222}$

13. $7\overline{)7076}$

14. $9\overline{)9098}$

15. $2\overline{)4471}$

Practice 18

Reminders

- A dividend is divisible by a divisor if it can be divided evenly by that divisor with no remainder.
- Any number ending in 0 or 5 is divisible by 5.
- Any number ending in 0 is divisible by 10.
- Any number ending in 00 is divisible by 100.

Directions: Do these problems. Use your multiplication/division chart if you are unsure of your division facts. The first one is done for you.

1.
```
       56
  10) 560
     - 50
       60
     - 60
        0
```

2. 10) 880

3. 10) 340

4. 5) 450

5. 5) 835

6. 5) 645

7. 100) 800

8. 100) 500

9. 100) 300

10. 10) 5660

11. 10) 9390

12. 10) 300

13. 5) 8755

14. 5) 7070

15. 5) 7080

16. 100) 4600

Practice 19

> ## Reminders
>
> - A dividend is divisible by a divisor if it can be divided evenly by that divisor with no remainder.
> - Any number ending in 0, 2, 4, 6, or 8 is divisible by 2.
> - A number is divisible by 4 if the last two digits are multiples of 4 such as 12, 16, 20, 24, 28, etc.

Directions: Do these problems. Use your multiplication/division chart if you are unsure of your division facts. The first one is done for you.

1.
```
      91
  2)182
  - 18
      2
    - 2
      0
```

2. 4)724

3. 2)628

4. 2)826

5. 4)928

6. 2)768

7. 2)518

8. 4)616

9. 2)924

10. 2)7324

11. 4)8744

12. 4)1992

13. 2)2826

14. 4)6716

15. 4)8984

16. 2)1718

Practice 20 ⟲ ⟲ ⟲ ⟲ ⟲ ⟲ ⟲ ⟲ ⟲ ⟲ ⟲ ⟲

> ## Reminders
> - A dividend is divisible by a divisor if it can be divided evenly by that divisor with no remainder.
> - A dividend is divisible by 3 if the sum of the digits in the dividend equals 3 or a multiple of 3 such as 6, 9, 12, 15, etc.
> - A dividend is divisible by 9 if the sum of the digits in the dividend equals 9 or a multiple of 9 such as 18, 27, 36, etc.

Directions: Do these problems. Use your multiplication/division chart if you are unsure of your division facts. The first one has been done for you.

1.
$$
\begin{array}{r}
53 \\
3\overline{)159} \\
-15 \\
\hline
9 \\
-9 \\
\hline
0
\end{array}
$$

2. $9\overline{)549}$

3. $3\overline{)633}$

4. $3\overline{)639}$

5. $9\overline{)459}$

6. $3\overline{)183}$

7. $9\overline{)540}$

8. $3\overline{)426}$

9. $3\overline{)375}$

10. $9\overline{)3222}$

11. $3\overline{)1236}$

12. $3\overline{)5166}$

13. $9\overline{)2754}$

14. $9\overline{)6471}$

15. $3\overline{)1224}$

16. $9\overline{)7488}$

Practice 21 ◐ ☙ ◐ ☙ ◐ ☙ ◐ ☙ ◐ ☙ ◐ ☙ ◐ ◐ ☙

Reminders

- A dividend is divisible by a divisor if it can be divided evenly by that divisor with no remainder.
- Any number ending in 00, 20, 40, 60, or 80 is divisible by 20.
- Any number ending in 00, 25, 50, or 75 is divisible by 25.

Directions: Do these problems. Use your multiplication/division chart if you are unsure of your division facts. The first one has been done for you.

1.
$$\begin{array}{r} 7 \\ 25\overline{)175} \\ -175 \\ \hline 0 \end{array}$$

2. $25\overline{)525}$

3. $25\overline{)425}$

4. $25\overline{)575}$

5. $25\overline{)925}$

6. $25\overline{)500}$

7. $20\overline{)400}$

8. $20\overline{)140}$

9. $20\overline{)180}$

10. $20\overline{)240}$

11. $20\overline{)880}$

12. $20\overline{)700}$

13. $25\overline{)3125}$

14. $25\overline{)1150}$

15. $25\overline{)8875}$

16. $20\overline{)8820}$

Practice 22

Directions: Do these problems. Use your multiplication/division chart if you are unsure of your division facts. The first one is done for you.

1. 20)140 − 140 = 0, quotient 7
2. 20)360
3. 20)480
4. 30)150
5. 30)240
6. 30)390
7. 40)480
8. 30)510
9. 30)960
10. 50)250
11. 20)980
12. 50)550
13. 20)460
14. 50)450
15. 50)850
16. 40)760

Practice 23 ㆍ ◎ ㆍ ◎ ㆍ ◎ ㆍ ◎ ㆍ ◎ ㆍ ◎ ㆍ ㆍ ◎

Directions: Do these problems. Use your multiplication/division chart if you are unsure of your division facts. The first one is done for you.

1.
$$30\overline{)270} \quad \begin{array}{r} 9 \\ \hline \end{array}$$
$$\underline{-270}$$
$$0$$

2. $50\overline{)450}$

3. $40\overline{)280}$

4. $70\overline{)280}$

5. $80\overline{)160}$

6. $60\overline{)360}$

7. $90\overline{)990}$

8. $40\overline{)360}$

9. $60\overline{)120}$

10. $80\overline{)640}$

11. $20\overline{)960}$

12. $70\overline{)210}$

13. $20\overline{)360}$

14. $70\overline{)420}$

15. $90\overline{)180}$

16. $50\overline{)750}$

Practice 24

Directions: Do these problems. Use your multiplication/division chart if you are unsure of your division facts. The first one is done for you.

1.
```
       4  R10
  40 ) 170
     - 160
       10
```

2. 20) 190

3. 60) 150

4. 60) 370

5. 50) 230

6. 40) 250

7. 70) 220

8. 90) 460

9. 40) 390

10. 70) 170

11. 60) 290

12. 70) 750

13. 30) 260

14. 70) 440

15. 20) 550

16. 70) 430

Practice 25

Directions: Do these problems. Use your multiplication/division chart if you are unsure of your division facts. The first one is done for you.

1.
```
      51 R10
50)2560
  - 250
     60
   - 50
     10
```

2. 40)3630

3. 20)1870

4. 40)3690

5. 30)2780

6. 70)2510

7. 90)2780

8. 60)4850

9. 40)8410

10. 70)3710

11. 80)2460

12. 90)7520

13. 40)2990

14. 60)9610

15. 20)8830

16. 70)4960

Practice 26 ᵔᵔᵔᵔᵔᵔᵔᵔᵔᵔᵔᵔᵔᵔ

Directions: Do these problems. Use your multiplication/division chart if you are unsure of your division facts. The first one has been done for you.

1.
```
        921
  30 )27,630
     -270
       63
      -60
       30
      -30
        0
```

2. $20 \overline{)48,400}$

3. $90 \overline{)18,990}$

4. $60 \overline{)66,600}$

5. $30 \overline{)90,900}$

6. $70 \overline{)14,700}$

7. $70 \overline{)28,700}$

8. $40 \overline{)44,400}$

9. $50 \overline{)85,500}$

10. $80 \overline{)24,080}$

11. $40 \overline{)52,080}$

12. $30 \overline{)96,060}$

13. $40 \overline{)32,690}$

14. $70 \overline{)87,130}$

15. $30 \overline{)70,710}$

Practice 27 ⟍ ◔ ⟍ ◔ ⟍ ◔ ⟍ ◔ ⟍ ◔ ⟍ ◔ ⟍ ⟍ ◔

Helpful Hint: When dividing by 25, think quarters.

Example: $25\overline{)88}$

Think: How many quarters could you have in 88 cents?

Answer: 3 quarters and 13 cents left over.

$$\begin{array}{r} 3 \text{ R13} \\ 25\overline{)88} \\ -75 \\ \hline 13 \end{array}$$

Directions: Do these problems. Use your multiplication/division chart if you are unsure of your division facts. The first one is done for you.

1.
$$\begin{array}{r} 6 \text{ R5} \\ 25\overline{)155} \\ -150 \\ \hline 5 \end{array}$$

2. $25\overline{)129}$

3. $25\overline{)240}$

4. $25\overline{)256}$

5. $25\overline{)828}$

6. $25\overline{)269}$

7. $25\overline{)309}$

8. $25\overline{)149}$

9. $25\overline{)188}$

10. $25\overline{)386}$

11. $25\overline{)445}$

12. $25\overline{)801}$

13. $25\overline{)8827}$

14. $25\overline{)3765}$

15. $25\overline{)9999}$

16. $25\overline{)7719}$

Practice 28

Directions: Do these problems. Use your multiplication/division chart if you are unsure of your division facts. The first one is done for you.

1.
```
     7 R7
22)161
  -154
     7
```

2. 22)254

3. 31)159

4. 23)131

5. 41)249

6. 21)498

7. 31)679

8. 32)995

9. 22)617

10. 19)418

11. 29)923

12. 52)759

13. 18)473

14. 42)893

15. 15)799

16. 37)816

Two-Digit Divisors (Not Ending in Zero)/Three-Digit Dividends

Practice 29 ⊘ ☙ ⊘ ☙ ⊘ ☙ ⊘ ☙ ⊘ ☙ ⊘ ☙ ⊘ ☙ ⊘

Directions: Do these problems. Use your multiplication/division chart if you are unsure of your division facts. The first one is done for you.

1.
$$
\begin{array}{r}
11\ \text{R}27 \\
31\overline{)368} \\
-31 \\
\hline
58 \\
-31 \\
\hline
27
\end{array}
$$

2. $24\overline{)499}$

3. $21\overline{)468}$

4. $22\overline{)297}$

5. $33\overline{)712}$

6. $23\overline{)691}$

7. $41\overline{)888}$

8. $51\overline{)616}$

9. $19\overline{)889}$

10. $24\overline{)509}$

11. $39\overline{)813}$

12. $42\overline{)919}$

13. $18\overline{)549}$

14. $49\overline{)537}$

15. $26\overline{)521}$

16. $47\overline{)967}$

Practice 30

Directions: Do these problems. Use your multiplication/division chart if you are unsure of your division facts. The first one is done for you.

1.
```
        203 R1
   21 )4264
      - 42
        64
      - 63
         1
```

2. 31)9363

3. 32)6466

4. 24)5026

5. 19)4080

6. 23)9913

7. 28)6023

8. 15)6048

9. 27)8418

10. 33)9999

11. 31)6362

12. 52)6224

13. 18)5455

14. 61)6612

15. 16)4850

Practice 31

Directions: Do these problems. Use your multiplication/division chart if you are unsure of your division facts. The first one is done for you.

1.
$$
\begin{array}{r}
201\ \text{R}28 \\
29\overline{)5857} \\
-58 \\
\hline
57 \\
-29 \\
\hline
28
\end{array}
$$

2. $38\overline{)7775}$

3. $28\overline{)5659}$

4. $38\overline{)8145}$

5. $16\overline{)5123}$

6. $29\overline{)9146}$

7. $34\overline{)6344}$

8. $19\overline{)5608}$

9. $37\overline{)7301}$

10. $13\overline{)5163}$

11. $28\overline{)5554}$

12. $57\overline{)5678}$

13. $28\overline{)2719}$

14. $42\overline{)1439}$

15. $27\overline{)2525}$

#3324 Practice Makes Perfect: Division

Practice 32

Directions: Do these problems. Use your multiplication/division chart if you are unsure of your division facts. The first one is done for you.

1.
$$
\begin{array}{r}
2 \text{ R35} \\
100\overline{)235} \\
-200 \\
\hline
35
\end{array}
$$

2. $100\overline{)457}$

3. $100\overline{)376}$

4. $200\overline{)800}$

5. $300\overline{)900}$

6. $300\overline{)600}$

7. $200\overline{)950}$

8. $400\overline{)870}$

9. $600\overline{)9980}$

10. $300\overline{)759}$

11. $400\overline{)965}$

12. $300\overline{)927}$

13. $100\overline{)2780}$

14. $100\overline{)1890}$

15. $100\overline{)3650}$

16. $200\overline{)9650}$

Practice 33

Directions: Do these problems. Don't forget to use dollar signs and decimal points. Use your multiplication/division chart if you are unsure of your division facts. The first one is done for you.

1.
```
        $.99
   5 )$4.95
     -4 5
       45
      -45
        0
```

2. 4)$2.84

3. 9)$6.39

4. 2)$8.80

5. 9)$3.24

6. 6)$9.18

7. 8)$9.60

8. 5)$8.90

9. 4)$2.36

10. 7)$8.47

11. 9)$9.36

12. 4)$8.92

13. 5)$88.35

14. 7)$77.14

15. 9)$88.11

16. 4)$21.88

Practice 34

Directions: Do these problems. Don't forget to use dollar signs and decimal points. Use your multiplication/division chart if you are unsure of your division facts. The first one is done for you.

1.
```
        $.21
   20 )$4.20
     - 4 0
        20
      - 20
         0
```

2. 30)$6.30

3. 40)$4.80

4. 20)$8.60

5. 30)$9.90

6. 40)$2.40

7. 60)$3.60

8. 50)$9.50

9. 20)$7.40

10. 25)$8.25

11. 25)$9.75

12. 25)$3.25

13. 25)$99.75

14. 50)$37.50

15. 75)$75.75

16. 21)$42.84

Simple Word Problems with Division

Practice 35

Directions: Use your division skills to solve these word problems. Use your multiplication/division chart, if needed.

1. What is 219 divided by 3? _____

2. Her teacher asked Christina to divide 1,230 straws among the 30 members of the class for a science experiment. How many straws did each student receive? _____

3. Divide 963 by 9. _____

4. Alyssa had $29.25 in her piggy bank. All of the money was in quarters (25¢). How many quarters did she have in her piggy bank? _____

5. The divisor is 29. The dividend is 986. What is the quotient? _____

6. What is the quotient when 1,024 is divided by 4? _____

7. Divide 2,340 by 20. _____

8. His science teacher asked James to pass out 510 milliliters of soap in 30 milliliter containers to each student. How many students received soap? _____

9. What is 1,333 divided by 31? _____

10. Find the quotient: 8,088 divided by 40. _____

11. Robert had to split a jar of 3,275 pieces of candy corn among 25 children at a Halloween party. How many candy corn pieces did each child receive? _____

12. What is 9,570 divided by 33? _____

Practice 36

Directions: Do these problems. Use your multiplication/division chart if you are unsure of your division facts. The first one is done for you.

1. $$250{\overline{\smash{\big)}\,750}}$$ with quotient 3 and -750 shown below

2. $$325{\overline{\smash{\big)}\,975}}$$

3. $$450{\overline{\smash{\big)}\,900}}$$

4. $$225{\overline{\smash{\big)}\,925}}$$

5. $$325{\overline{\smash{\big)}\,875}}$$

6. $$255{\overline{\smash{\big)}\,595}}$$

7. $$425{\overline{\smash{\big)}\,875}}$$

8. $$325{\overline{\smash{\big)}\,770}}$$

9. $$625{\overline{\smash{\big)}\,995}}$$

10. $$245{\overline{\smash{\big)}\,785}}$$

11. $$125{\overline{\smash{\big)}\,855}}$$

12. $$175{\overline{\smash{\big)}\,780}}$$

13. $$250{\overline{\smash{\big)}\,2250}}$$

14. $$125{\overline{\smash{\big)}\,1375}}$$

15. $$350{\overline{\smash{\big)}\,7350}}$$

16. $$550{\overline{\smash{\big)}\,8500}}$$

Test Practice 1

1. $5\overline{)95}$
    - Ⓐ 21
    - Ⓑ 18
    - Ⓒ 19
    - Ⓓ 20

2. $7\overline{)91}$
    - Ⓐ 13
    - Ⓑ 23
    - Ⓒ 15
    - Ⓓ 17

3. $6\overline{)96}$
    - Ⓐ 17
    - Ⓑ 18
    - Ⓒ 16
    - Ⓓ 19

4. $3\overline{)87}$
    - Ⓐ 19
    - Ⓑ 39
    - Ⓒ 27
    - Ⓓ 29

5. $6\overline{)84}$
    - Ⓐ 15
    - Ⓑ 17
    - Ⓒ 24
    - Ⓓ 14

6. $7\overline{)98}$
    - Ⓐ 14
    - Ⓑ 19
    - Ⓒ 13
    - Ⓓ 15

7. $4\overline{)192}$
    - Ⓐ 48
    - Ⓑ 47
    - Ⓒ 58
    - Ⓓ 49

8. $5\overline{)435}$
    - Ⓐ 88
    - Ⓑ 86
    - Ⓒ 97
    - Ⓓ 87

9. $6\overline{)834}$
    - Ⓐ 149
    - Ⓑ 139
    - Ⓒ 138
    - Ⓓ 141

10. $8\overline{)664}$
    - Ⓐ 84
    - Ⓑ 88
    - Ⓒ 82
    - Ⓓ 83

11. $3\overline{)831}$
    - Ⓐ 257
    - Ⓑ 287
    - Ⓒ 277
    - Ⓓ 267

12. $6\overline{)438}$
    - Ⓐ 83
    - Ⓑ 76
    - Ⓒ 74
    - Ⓓ 73

13. $9\overline{)243}$
    - Ⓐ 29
    - Ⓑ 27
    - Ⓒ 28
    - Ⓓ 37

14. $7\overline{)917}$
    - Ⓐ 133
    - Ⓑ 137
    - Ⓒ 121
    - Ⓓ 131

Test Practice 2

1. 6)129
   - Ⓐ 21 R3
   - Ⓑ 22 R3
   - Ⓒ 20 R3
   - Ⓓ 21 R1

2. 4)417
   - Ⓐ 14 R1
   - Ⓑ 104 R3
   - Ⓒ 106 R1
   - Ⓓ 104 R1

3. 9)643
   - Ⓐ 70 R4
   - Ⓑ 72 R4
   - Ⓒ 71 R4
   - Ⓓ 73 R4

4. 7)818
   - Ⓐ 116 R1
   - Ⓑ 116 R6
   - Ⓒ 117 R6
   - Ⓓ 116 R5

5. 4)197
   - Ⓐ 48 R3
   - Ⓑ 49 R1
   - Ⓒ 47 R1
   - Ⓓ 49 R3

6. 5)983
   - Ⓐ 196 R4
   - Ⓑ 195 R3
   - Ⓒ 196 R3
   - Ⓓ 186 R3

7. 7)159
   - Ⓐ 22 R5
   - Ⓑ 23 R5
   - Ⓒ 22 R4
   - Ⓓ 24 R5

8. 5)478
   - Ⓐ 95 R4
   - Ⓑ 95 R3
   - Ⓒ 96 R3
   - Ⓓ 96 R4

9. 8)659
   - Ⓐ 82 R5
   - Ⓑ 82 R7
   - Ⓒ 92 R3
   - Ⓓ 82 R3

10. 6)737
    - Ⓐ 123 R5
    - Ⓑ 125 R3
    - Ⓒ 122 R5
    - Ⓓ 124 R5

11. 9)862
    - Ⓐ 94 R7
    - Ⓑ 96 R7
    - Ⓒ 95 R7
    - Ⓓ 95 R6

12. 3)796
    - Ⓐ 265 R2
    - Ⓑ 265 R1
    - Ⓒ 264 R2
    - Ⓓ 266 R1

13. 7)325
    - Ⓐ 47 R3
    - Ⓑ 46 R2
    - Ⓒ 48 R3
    - Ⓓ 46 R3

14. 6)831
    - Ⓐ 137 R4
    - Ⓑ 137 R3
    - Ⓒ 138 R2
    - Ⓓ 138 R3

Test Practice 3

1. $5\overline{)105}$
   - Ⓐ 37
   - Ⓑ 38
   - Ⓒ 49
   - Ⓓ 21

2. $10\overline{)830}$
   - Ⓐ 83
   - Ⓑ 830
   - Ⓒ 84
   - Ⓓ 803

3. $5\overline{)795}$
   - Ⓐ 157
   - Ⓑ 167
   - Ⓒ 15
   - Ⓓ 159

4. $25\overline{)875}$
   - Ⓐ 35
   - Ⓑ 335
   - Ⓒ 36
   - Ⓓ 25

5. $4\overline{)256}$
   - Ⓐ 65
   - Ⓑ 64
   - Ⓒ 67
   - Ⓓ 66

6. $2\overline{)322}$
   - Ⓐ 163
   - Ⓑ 161
   - Ⓒ 261
   - Ⓓ 163

7. $3\overline{)162}$
   - Ⓐ 55
   - Ⓑ 57
   - Ⓒ 54
   - Ⓓ 56

8. $9\overline{)657}$
   - Ⓐ 73
   - Ⓑ 63
   - Ⓒ 72
   - Ⓓ 74

9. $3\overline{)126}$
   - Ⓐ 44
   - Ⓑ 43
   - Ⓒ 52
   - Ⓓ 42

10. $3\overline{)423}$
    - Ⓐ 142
    - Ⓑ 141
    - Ⓒ 151
    - Ⓓ 143

11. $5\overline{)805}$
    - Ⓐ 162
    - Ⓑ 163
    - Ⓒ 160
    - Ⓓ 161

12. $20\overline{)940}$
    - Ⓐ 407
    - Ⓑ 46
    - Ⓒ 47
    - Ⓓ 417

13. $2\overline{)866}$
    - Ⓐ 443
    - Ⓑ 423
    - Ⓒ 432
    - Ⓓ 433

14. $10\overline{)630}$
    - Ⓐ 630
    - Ⓑ 63
    - Ⓒ 603
    - Ⓓ 613

#3324 Practice Makes Perfect: Division

Test Practice 4

1. $30{\overline{)180}}$
   - Ⓐ 7
   - Ⓑ 8
   - Ⓒ 9
   - Ⓓ 6

2. $20{\overline{)480}}$
   - Ⓐ 23
   - Ⓑ 25
   - Ⓒ 13
   - Ⓓ 24

3. $50{\overline{)350}}$
   - Ⓐ 6
   - Ⓑ 9
   - Ⓒ 70
   - Ⓓ 7

4. $70{\overline{)630}}$
   - Ⓐ 9
   - Ⓑ 90
   - Ⓒ 8
   - Ⓓ 7

5. $40{\overline{)840}}$
   - Ⓐ 31
   - Ⓑ 22
   - Ⓒ 21
   - Ⓓ 22

6. $60{\overline{)960}}$
   - Ⓐ 16
   - Ⓑ 18
   - Ⓒ 26
   - Ⓓ 17

7. $30{\overline{)170}}$
   - Ⓐ 4 R20
   - Ⓑ 5 R10
   - Ⓒ 5 R20
   - Ⓓ 6 R20

8. $40{\overline{)910}}$
   - Ⓐ 21 R30
   - Ⓑ 220 R30
   - Ⓒ 22 R20
   - Ⓓ 22 R30

9. $60{\overline{)770}}$
   - Ⓐ 13 R40
   - Ⓑ 12 R20
   - Ⓒ 12 R30
   - Ⓓ 12 R50

10. $70{\overline{)950}}$
    - Ⓐ 12 R40
    - Ⓑ 13 R20
    - Ⓒ 13 R50
    - Ⓓ 13 R40

11. $80{\overline{)250}}$
    - Ⓐ 3 R10
    - Ⓑ 4 R10
    - Ⓒ 31 R10
    - Ⓓ 3 R20

12. $50{\overline{)475}}$
    - Ⓐ 9 R15
    - Ⓑ 8 R15
    - Ⓒ 9 R25
    - Ⓓ 9 R40

13. $60{\overline{)904}}$
    - Ⓐ 15 R4
    - Ⓑ 16 R4
    - Ⓒ 14 R4
    - Ⓓ 25 R4

14. $30{\overline{)736}}$
    - Ⓐ 25 R19
    - Ⓑ 24 R19
    - Ⓒ 24 R16
    - Ⓓ 25 R16

Test Practice 5

1. $25\overline{)225}$
   - Ⓐ 7
   - Ⓑ 8
   - Ⓒ 9
   - Ⓓ 6

2. $25\overline{)575}$
   - Ⓐ 23
   - Ⓑ 25
   - Ⓒ 13
   - Ⓓ 24

3. $35\overline{)770}$
   - Ⓐ 22
   - Ⓑ 20
   - Ⓒ 21
   - Ⓓ 31

4. $22\overline{)660}$
   - Ⓐ 20
   - Ⓑ 30
   - Ⓒ 21
   - Ⓓ 25

5. $33\overline{)990}$
   - Ⓐ 31
   - Ⓑ 32
   - Ⓒ 30
   - Ⓓ 21

6. $19\overline{)760}$
   - Ⓐ 41
   - Ⓑ 42
   - Ⓒ 39
   - Ⓓ 40

7. $31\overline{)197}$
   - Ⓐ 6 R13
   - Ⓑ 6 R9
   - Ⓒ 5 R11
   - Ⓓ 6 R11

8. $29\overline{)623}$
   - Ⓐ 21 R13
   - Ⓑ 21 R12
   - Ⓒ 22 R14
   - Ⓓ 21 R14

9. $27\overline{)611}$
   - Ⓐ 22 R17
   - Ⓑ 12 R17
   - Ⓒ 22 R16
   - Ⓓ 22 R18

10. $29\overline{)765}$
    - Ⓐ 26 R11
    - Ⓑ 27 R13
    - Ⓒ 26 R9
    - Ⓓ 26 R12

11. $24\overline{)6814}$
    - Ⓐ 283 R23
    - Ⓑ 293 R22
    - Ⓒ 273 R24
    - Ⓓ 283 R22

12. $33\overline{)9745}$
    - Ⓐ 294 R15
    - Ⓑ 295 R15
    - Ⓒ 295 R10
    - Ⓓ 296 R10

13. $14\overline{)6109}$
    - Ⓐ 437 R5
    - Ⓑ 436 R15
    - Ⓒ 435 R5
    - Ⓓ 436 R5

14. $31\overline{)9432}$
    - Ⓐ 304 R18
    - Ⓑ 304 R8
    - Ⓒ 306 R8
    - Ⓓ 306 R18

Test Practice 6 ⊙ ☽ ⊙ ☽ ⊙ ☽ ⊙ ☽ ⊙ ☽ ⊙ ☽

1. $6\overline{)546}$
   - Ⓐ 92
   - Ⓑ 91
   - Ⓒ 81
   - Ⓓ 93

8. $4\overline{)7313}$
   - Ⓐ 1829 R1
   - Ⓑ 1827 R3
   - Ⓒ 1828 R3
   - Ⓓ 1828 R1

2. $9\overline{)720}$
   - Ⓐ 80
   - Ⓑ 81
   - Ⓒ 60
   - Ⓓ 90

9. $25\overline{)487}$
   - Ⓐ 18 R17
   - Ⓑ 19 R18
   - Ⓒ 19 R12
   - Ⓓ 19 R13

3. $9\overline{)837}$
   - Ⓐ 94
   - Ⓑ 93
   - Ⓒ 91
   - Ⓓ 97

10. $30\overline{)689}$
   - Ⓐ 22 R19
   - Ⓑ 23 R29
   - Ⓒ 24 R29
   - Ⓓ 22 R29

4. $7\overline{)434}$
   - Ⓐ 63
   - Ⓑ 66
   - Ⓒ 72
   - Ⓓ 62

11. $38\overline{)887}$
   - Ⓐ 23 R13
   - Ⓑ 23 R14
   - Ⓒ 24 R12
   - Ⓓ 23 R17

5. $6\overline{)197}$
   - Ⓐ 32 R5
   - Ⓑ 32 R3
   - Ⓒ 33 R3
   - Ⓓ 22 R3

12. $29\overline{)764}$
   - Ⓐ 25 R10
   - Ⓑ 26 R20
   - Ⓒ 26 R11
   - Ⓓ 26 R10

6. $8\overline{)617}$
   - Ⓐ 76 R1
   - Ⓑ 77 R3
   - Ⓒ 75 R1
   - Ⓓ 77 R1

13. $23\overline{)8811}$
   - Ⓐ 383 R3
   - Ⓑ 383 R2
   - Ⓒ 373 R2
   - Ⓓ 383 R4

7. $5\overline{)2733}$
   - Ⓐ 546 R4
   - Ⓑ 545 R4
   - Ⓒ 546 R3
   - Ⓓ 536 R3

14. $39\overline{)9856}$
   - Ⓐ 253 R38
   - Ⓑ 252 R38
   - Ⓒ 252 R18
   - Ⓓ 252 R28

Answer Sheet

Test Practice 1	Test Practice 2	Test Practice 3
1. Ⓐ Ⓑ Ⓒ Ⓓ	1. Ⓐ Ⓑ Ⓒ Ⓓ	1. Ⓐ Ⓑ Ⓒ Ⓓ
2. Ⓐ Ⓑ Ⓒ Ⓓ	2. Ⓐ Ⓑ Ⓒ Ⓓ	2. Ⓐ Ⓑ Ⓒ Ⓓ
3. Ⓐ Ⓑ Ⓒ Ⓓ	3. Ⓐ Ⓑ Ⓒ Ⓓ	3. Ⓐ Ⓑ Ⓒ Ⓓ
4. Ⓐ Ⓑ Ⓒ Ⓓ	4. Ⓐ Ⓑ Ⓒ Ⓓ	4. Ⓐ Ⓑ Ⓒ Ⓓ
5. Ⓐ Ⓑ Ⓒ Ⓓ	5. Ⓐ Ⓑ Ⓒ Ⓓ	5. Ⓐ Ⓑ Ⓒ Ⓓ
6. Ⓐ Ⓑ Ⓒ Ⓓ	6. Ⓐ Ⓑ Ⓒ Ⓓ	6. Ⓐ Ⓑ Ⓒ Ⓓ
7. Ⓐ Ⓑ Ⓒ Ⓓ	7. Ⓐ Ⓑ Ⓒ Ⓓ	7. Ⓐ Ⓑ Ⓒ Ⓓ
8. Ⓐ Ⓑ Ⓒ Ⓓ	8. Ⓐ Ⓑ Ⓒ Ⓓ	8. Ⓐ Ⓑ Ⓒ Ⓓ
9. Ⓐ Ⓑ Ⓒ Ⓓ	9. Ⓐ Ⓑ Ⓒ Ⓓ	9. Ⓐ Ⓑ Ⓒ Ⓓ
10. Ⓐ Ⓑ Ⓒ Ⓓ	10. Ⓐ Ⓑ Ⓒ Ⓓ	10. Ⓐ Ⓑ Ⓒ Ⓓ
11. Ⓐ Ⓑ Ⓒ Ⓓ	11. Ⓐ Ⓑ Ⓒ Ⓓ	11. Ⓐ Ⓑ Ⓒ Ⓓ
12. Ⓐ Ⓑ Ⓒ Ⓓ	12. Ⓐ Ⓑ Ⓒ Ⓓ	12. Ⓐ Ⓑ Ⓒ Ⓓ
13. Ⓐ Ⓑ Ⓒ Ⓓ	13. Ⓐ Ⓑ Ⓒ Ⓓ	13. Ⓐ Ⓑ Ⓒ Ⓓ
14. Ⓐ Ⓑ Ⓒ Ⓓ	14. Ⓐ Ⓑ Ⓒ Ⓓ	14. Ⓐ Ⓑ Ⓒ Ⓓ

Test Practice 4	Test Practice 5	Test Practice 6
1. Ⓐ Ⓑ Ⓒ Ⓓ	1. Ⓐ Ⓑ Ⓒ Ⓓ	1. Ⓐ Ⓑ Ⓒ Ⓓ
2. Ⓐ Ⓑ Ⓒ Ⓓ	2. Ⓐ Ⓑ Ⓒ Ⓓ	2. Ⓐ Ⓑ Ⓒ Ⓓ
3. Ⓐ Ⓑ Ⓒ Ⓓ	3. Ⓐ Ⓑ Ⓒ Ⓓ	3. Ⓐ Ⓑ Ⓒ Ⓓ
4. Ⓐ Ⓑ Ⓒ Ⓓ	4. Ⓐ Ⓑ Ⓒ Ⓓ	4. Ⓐ Ⓑ Ⓒ Ⓓ
5. Ⓐ Ⓑ Ⓒ Ⓓ	5. Ⓐ Ⓑ Ⓒ Ⓓ	5. Ⓐ Ⓑ Ⓒ Ⓓ
6. Ⓐ Ⓑ Ⓒ Ⓓ	6. Ⓐ Ⓑ Ⓒ Ⓓ	6. Ⓐ Ⓑ Ⓒ Ⓓ
7. Ⓐ Ⓑ Ⓒ Ⓓ	7. Ⓐ Ⓑ Ⓒ Ⓓ	7. Ⓐ Ⓑ Ⓒ Ⓓ
8. Ⓐ Ⓑ Ⓒ Ⓓ	8. Ⓐ Ⓑ Ⓒ Ⓓ	8. Ⓐ Ⓑ Ⓒ Ⓓ
9. Ⓐ Ⓑ Ⓒ Ⓓ	9. Ⓐ Ⓑ Ⓒ Ⓓ	9. Ⓐ Ⓑ Ⓒ Ⓓ
10. Ⓐ Ⓑ Ⓒ Ⓓ	10. Ⓐ Ⓑ Ⓒ Ⓓ	10. Ⓐ Ⓑ Ⓒ Ⓓ
11. Ⓐ Ⓑ Ⓒ Ⓓ	11. Ⓐ Ⓑ Ⓒ Ⓓ	11. Ⓐ Ⓑ Ⓒ Ⓓ
12. Ⓐ Ⓑ Ⓒ Ⓓ	12. Ⓐ Ⓑ Ⓒ Ⓓ	12. Ⓐ Ⓑ Ⓒ Ⓓ
13. Ⓐ Ⓑ Ⓒ Ⓓ	13. Ⓐ Ⓑ Ⓒ Ⓓ	13. Ⓐ Ⓑ Ⓒ Ⓓ
14. Ⓐ Ⓑ Ⓒ Ⓓ	14. Ⓐ Ⓑ Ⓒ Ⓓ	14. Ⓐ Ⓑ Ⓒ Ⓓ

Answer Key

Page 4
1. 12, 10, 8, 6, 4, 2
2. 18, 15, 12, 9, 6, 3
3. 32, 28, 24, 20, 16, 12, 8, 4
4. 35, 30, 25, 20, 15, 10, 5
5. 42, 36, 30, 24, 18, 12, 6
6. 49, 42, 35, 28, 21, 14, 7
7. 56, 48, 40, 32, 24, 16, 8
8. 72, 60, 48, 36, 24, 12
9. the 10's

Page 5
1. 56
48
40
32
24
16
8
2. 70
60
50
40
30
20
10
3. 35
30
25
20
15
10
5
4. 77
66
55
44
33
22
11
5. 84
72
60
48
36
24
12
6. ten's (#2)
7. 6
8. 6
9. 8
10. 6
11. 12
12. 10

Page 6
1. 5
2. 5
3. 3
4. 1
5. 9
6. 5
7. 9
8. 7
9. 5
10. 8
11. 1
12. 10
13. 11
14. 9
15. 6
16. 8
17. 5
18. 12
19. 3
20. 4
21. 6
22. 6
23. 4
24. 6
25. 7
26. 3
27. 7
28. 10
29. 8
30. 9
31. 11
32. 12
33. 9
34. 12
35. 12
36. 11
37. 10
38. 6
39. 10
40. 9
41. 12
42. 7
43. 11
44. 10
45. 8
46. 10
47. 8
48. 11

Page 7
1. 4
2. 7
3. 3
4. 6
5. 8
6. 9
7. 6
8. 8
9. 8
10. 5
11. 3
12. 5
13. 2
14. 8
15. 3
16. 3
17. 11
18. 11
19. 6
20. 1
21. 7
22. 3
23. 10
24. 12
25. 2
26. 3
27. 9
28. 10
29. 3
30. 11
31. 4
32. 4
33. 4
34. 1
35. 9
36. 9
37. 5
38. 9
39. 10
40. 2
41. 11
42. 12
43. 7
44. 7
45. 2
46. 9
47. 12
48. 12

Page 8
1. 1
2. 5
3. 5
4. 4
5. 2
6. 11
7. 8
8. 8
9. 6
10. 9
11. 8
12. 10
13. 6
14. 4
15. 2
16. 12
17. 5
18. 3
19. 7
20. 12
21. 3
22. 4
23. 9
24. 10
25. 5
26. 3
27. 2
28. 3
29. 8
30. 11
31. 6
32. 3
33. 1
34. 4
35. 6
36. 11
37. 5
38. 6
39. 1
40. 12
41. 10
42. 9
43. 11
44. 7
45. 4
46. 5
47. 5
48. 9

Page 9
1. 3
2. 6
3. 8
4. 4
5. 8
6. 3
7. 12
8. 5
9. 8
10. 5
11. 7
12. 9
13. 4
14. 9
15. 4
16. 3
17. 11
18. 8
19. 11
20. 11
21. 6
22. 2
23. 4
24. 5
25. 10
26. 4
27. 8
28. 3
29. 5
30. 5
31. 3
32. 12
33. 5
34. 12
35. 9
36. 7
37. 7
38. 5
39. 5
40. 10
41. 2
42. 10
43. 8
44. 12
45. 5
46. 3
47. 9
48. 7

Page 10
1. 7
2. 5
3. 3
4. 8
5. 7
6. 3
7. 9
8. 5
9. 8
10. 6
11. 5
12. 2
13. 12
14. 3
15. 4
16. 4
17. 7
18. 8
19. 9
20. 10
21. 6
22. 4
23. 9
24. 11
25. 3
26. 3
27. 11
28. 7
29. 7
30. 10
31. 11
32. 3
33. 9
34. 8
35. 9
36. 7
37. 5
38. 10
39. 4
40. 5
41. 6
42. 9
43. 12
44. 12
45. 5
46. 9
47. 2
48. 7

Page 11
1. 6
2. 12
3. 6
4. 8
5. 9
6. 12
7. 5
8. 12
9. 6
10. 9
11. 7
12. 11
13. 6
14. 9
15. 9
16. 11
17. 7
18. 7
19. 6
20. 6
21. 9
22. 4
23. 11
24. 4
25. 8
26. 7
27. 12
28. 3
29. 11
30. 9
31. 11
32. 12
33. 7
34. 6
35. 7
36. 8

Page 12
1. 9 R3
2. 9 R3
3. 12 R1
4. 11 R1
5. 5 R2
6. 11 R1
7. 6 R4
8. 11 R7
9. 7 R5
10. 8 R5
11. 12 R3
12. 5 R6
13. 6 R5
14. 21 R1
15. 17 R2
16. 12 R1
17. 12 R1
18. 14 R3
19. 19 R4
20. 15 R1
21. 13 R1
22. 15 R1
23. 22 R3
24. 24 R1

Page 13
1. 11 R4
2. 8 R4
3. 9 R3
4. 13 R1
5. 9 R1
6. 7 R4
7. 13 R4
8. 4 R7
9. 6 R3
10. 10 R7
11. 15 R2
12. 5 R9
13. 7 R8
14. 9 R2
15. 12 R3
16. 15 R2
17. 11 R5
18. 22 R1
19. 8 R3
20. 6 R7
21. 11 R5
22. 13 R2
23. 8 R6
24. 15 R3

Page 14
1. 51
2. 31
3. 41
4. 81
5. 211
6. 71
7. 71
8. 41
9. 91
10. 62
11. 222
12. 61
13. 62
14. 52
15. 75
16. 86
17. 34
18. 53
19. 75
20. 99

Page 15
1. 92
2. 123
3. 216
4. 128
5. 137
6. 74
7. 142
8. 246
9. 252
10. 234
11. 163
12. 101
13. 120
14. 206
15. 106

Page 16
1. 103 R3
2. 77 R1
3. 53 R4
4. 106 R1
5. 152 R5
6. 29 R2
7. 13 R5
8. 163 R1
9. 101 R2
10. 104 R3
11. 134 R2
12. 204 R2
13. 228 R3
14. 182 R1
15. 309 R2
16. 190 R1

Page 17
1. 122 R5
2. 175 R2
3. 73 R1
4. 132 R3
5. 123 R2
6. 96 R1
7. 72 R5
8. 29 R7
9. 214 R1
10. 45 R3
11. 104 R1
12. 89 R6
13. 227 R1
14. 244 R1
15. 198 R1
16. 30 R5

Page 18
1. 716
2. 515
3. 803
4. 804
5. 603
6. 607
7. 510
8. 1101
9. 801
10. 841
11. 814

Answer Key

12. 1647
13. 1382
14. 358
15. 1101
Page 19
1. 590 R3
2. 717 R1
3. 552 R1
4. 642 R1
5. 829 R1
6. 546 R6
7. 752 R3
8. 1188 R5
9. 843 R3
10. 943 R2
11. 979 R8
12. 729 R1
13. 213 R2
14. 333 R2
15. 1035 R1
Page 20
1. 645 R2
2. 1300 R3
3. 319 R5
4. 770 R4
5. 865 R2
6. 1498 R3
7. 327 R4
8. 1097 R2
9. 400 R2
10. 300 R7
11. 223 R1
12. 740 R2
13. 1010 R6
14. 1010 R8
15. 2235 R1
Page 21
1. 56
2. 88
3. 34
4. 90
5. 167
6. 129
7. 8
8. 5
9. 3
10. 566
11. 939
12. 30
13. 1751
14. 1414
15. 1416
16. 46
Page 22
1. 91
2. 181
3. 314
4. 413
5. 232
6. 384
7. 259
8. 154
9. 462
10. 3662

11. 2186
12. 498
13. 1413
14. 1679
15. 2246
16. 859
Page 23
1. 53
2. 61
3. 211
4. 213
5. 51
6. 61
7. 60
8. 142
9. 125
10. 358
11. 412
12. 1722
13. 306
14. 719
15. 408
16. 832
Page 24
1. 7
2. 21
3. 17
4. 23
5. 37
6. 20
7. 20
8. 7
9. 9
10. 12
11. 44
12. 35
13. 125
14. 46
15. 355
16. 441
Page 25
1. 7
2. 18
3. 24
4. 5
5. 8
6. 13
7. 12
8. 17
9. 32
10. 5
11. 49
12. 11
13. 23
14. 9
15. 17
16. 19
Page 26
1. 9
2. 9
3. 7
4. 4
5. 2
6. 6

7. 11
8. 9
9. 2
10. 8
11. 48
12. 3
13. 18
14. 6
15. 2
16. 15
Page 27
1. 4 R10
2. 9 R10
3. 2 R30
4. 6 R10
5. 4 R30
6. 6 R10
7. 3 R10
8. 5 R10
9. 9 R30
10. 2 R30
11. 4 R50
12. 10 R50
13. 8 R20
14. 6 R20
15. 27 R10
16. 6 R10
Page 28
1. 51 R10
2. 90 R30
3. 93 R10
4. 92 R10
5. 92 R20
6. 35 R60
7. 30 R80
8. 80 R50
9. 210 R10
10. 53
11. 30 R60
12. 83 R50
13. 74 R30
14. 160 R10
15. 441 R10
16. 70 R60
Page 29
1. 921
2. 2420
3. 211
4. 1110
5. 3030
6. 210
7. 410
8. 1110
9. 1710
10. 301
11. 1302
12. 3202
13. 817 R10
14. 1244
    R50
15. 2357
Page 30
1. 6 R5
2. 5 R4

3. 9 R15
4. 10 R6
5. 33 R3
6. 10 R19
7. 12 R9
8. 5 R24
9. 7 R13
10. 15 R11
11. 17 R20
12. 32 R1
13. 353 R2
14. 150 R15
15. 399 R24
16. 308 R19
Page 31
1. 7 R7
2. 11 R12
3. 5 R4
4. 5 R16
5. 6 R3
6. 23 R15
7. 21 R28
8. 31 R3
9. 28 R1
10. 22
11. 31 R24
12. 14 R31
13. 26 R5
14. 21 R11
15. 53 R4
16. 22 R2
Page 32
1. 11 R27
2. 20 R19
3. 22 R6
4. 13 R11
5. 21 R19
6. 30 R1
7. 21 R27
8. 12 R4
9. 46 R15
10. 21 R5
11. 20 R33
12. 21 R37
13. 30 R9
14. 10 R47
15. 20 R1
16. 20 R27
Page 33
1. 203 R1
2. 302 R1
3. 202 R2
4. 209 R10
5. 214 R14
6. 431
7. 215 R3
8. 403 R3
9. 311 R21
10. 303
11. 205 R7
12. 119 R36
13. 303 R1
14. 108 R24
15. 303 R2

Page 34
1. 201 R28
2. 204 R23
3. 202 R3
4. 214 R13
5. 320 R3
6. 315 R11
7. 186 R20
8. 295 R3
9. 197 R12
10. 397 R2
11. 198 R10
12. 99 R35
13. 97 R3
14. 34 R11
15. 93 R14
Page 35
1. 2 R35
2. 4 R57
3. 3 R76
4. 4
5. 3
6. 2
7. 4 R150
8. 2 R70
9. 16 R380
10. 2 R159
11. 2 R165
12. 3 R27
13. 27 R80
14. 18 R90
15. 36 R50
16. 48 R50
Page 36
1. $.99
2. $.71
3. $.71
4. $4.40
5. $.36
6. $1.53
7. $1.20
8. $1.78
9. $.59
10. $1.21
11. $1.04
12. $2.23
13. $17.67
14. $11.02
15. $9.79
16. $5.47
Page 37
1. $0.21
2. $0.21
3. $0.12
4. $0.43
5. $0.33
6. $0.06
7. $0.06
8. $0.19
9. $0.37
10. $0.33
11. $0.39
12. $0.13
13. $3.99

14. $0.75
15. $1.01
16. $2.04
Page 38
1. 73
2. 41 straws
3. 107
4. 117 quarters
5. 34
6. 256
7. 117
8. 17 students
9. 43
10. 202 R8
11. 131 candy
    corns
12. 290
Page 39
1. 3
2. 3
3. 2
4. 4 R25
5. 2 R225
6. 2 R85
7. 2 R25
8. 2 R120
9. 1 R370
10. 3 R50
11. 6 R105
12. 4 R80
13. 9
14. 11
15. 21
16. 15 R250
Page 40
1. C
2. A
3. C
4. D
5. D
6. A
7. A
8. D
9. B
10. D
11. C
12. D
13. B
14. D
Page 41
1. A
2. D
3. C
4. B
5. B
6. C
7. A
8. B
9. D
10. C
11. C
12. B
13. D
14. D

Page 42
1. D
2. A
3. D
4. A
5. B
6. B
7. C
8. A
9. D
10. B
11. D
12. C
13. D
14. B
Page 43
1. D
2. D
3. D
4. A
5. C
6. A
7. C
8. D
9. D
10. D
11. A
12. C
13. A
14. C
Page 44
1. C
2. A
3. A
4. B
5. C
6. D
7. D
8. D
9. A
10. A
11. D
12. C
13. D
14. B
Page 45
1. B
2. A
3. B
4. D
5. A
6. D
7. C
8. D
9. C
10. D
11. A
12. D
13. B
14. D